This book is not about COVID per se.
It's a fictional story about a family and city dealing with a ferocious "flu" pandemic. The influenza (Flu) and COVID-19 are both contagious respiratory illness, but they are caused by different viruses. COVID-19 is caused by infection with a new coronavirus (called SARS-CoV-2) and a flu is caused by infection with influenza viruses.

This book is dedicated to everyone affected in any way by the recent COVID pandemic.

SethSantoro.com
Smile From The Inside, Inc.

Thanks for my loved ones who motivate and inspire me to do better and be better every day.

A special thanks to my husband for his graphic design eye and his patience during this time of quarantine.

A special thanks to Fanie, Cari, Chris B, and Suzanne for their thoughts.

Dear Reader,

Finn & The Ferocious Flu is about how a family navigates through an unprecedented event in modern times.

This book is designed for your edutainment. In other words, the story within this book is meant to be both educational and entertaining.

As you read through the story, pay special attention to the family's emotions, feelings, and behaviors as they deal with this challenging time in their lives.

Afterwards, you will be prompted with questions and guided to discuss the various events and steps we all adventure through when faced with major loss and/or obstacles in your own life.

Hopefully, Finn's story will help you the next time you experience a significant loss in your life.

Thank you,

Contents

The **S.M.I.L.E.** Adventure

Finn
Mom
Finn and the Ferocious Flu
Brad
Dad

S

Once upon a time
in a city far away,
there lived a family, my family,
who was quite normal, I'd say.

Mom and Dad both worked hard,
my brother, Brad, a pain in the butt.
Just like normal brothers,
he didn't like me much, so what?

I was an eight-year-old kid
who just wanted to read books;
Play video games, have play dates,
futz with my dragon named Schnooks.

Mom was always caring
while Dad was otherwise occupied;
My brother was super silly, and
I always looked on the bright side.

Until one cold winter day,
a terrible illness spread wide;
with strength and determination,
a thorn in our side.

To flatten the flu's curve,
the City went into quarantine;
social distancing, the malls even closed,
something like we'd never seen.

Then poor dear Aunty Dede
she became ill with this flu;
Leaving this world so fast,
we all didn't know what to do.

And then everything changed,
Mom & Dad went into shock;
and forbade both Brad and I
from playing on our block.

S

Today it's been 2 weeks,
since stay at home orders passed;
And we have no idea what's next,
or how long this might last?

Mom started working from home
in our study every day;
New rule is she's not to be disturbed,
for now that's okay.

My parents are super-stressed
keeping us safe from what's beyond the door;
I do my best, give them space,
then, at bed time, love on them more.

Dad now often seems a bit confused,
like walking around in a daze;
But I'm loving every second,
when he sits down with us and plays.

S

Brad is still a dork
but keeps mostly to himself;
On his phone, the computer or
staring at his unique bookshelf.

Every once in a while,
he graces us with snarky 'tude;
So I annoy him a bit,
until he farts in my face, how rude!

Schools have been closed,
for what seems like a year.
Some people are going crazy,
while others are living in fear.

School now has these video calls,
I don't like them, not one bit.
What does it matter if I learn now,
why can't I just quit?

Today it's been four weeks,
and life is completely on hold.

I caught Mom crying last week,
thought it best not to interfere;
She even yelled at me for nothing,
for which I'm still a bit unclear.

Mom and Dad fight more and more,
and I dare **NOT** get involved;
I keep hoping and praying daily,
this will all soon be resolved.

Dad is doing his best,
more loving and chatty than usual;
He hugs me now quite often,
I don't mind, the feeling is mutual.

My favorite meals are breakfasts,
Dad makes me whatever I please.
He says this won't last forever!
But I secretly want time to freeze.

Dad has now taken charge
of food shopping and the like;
'cause Mom's too nervous to leave,
we joke around that she's on strike!

Dad doesn't like to go either,
to wait in those six-foot spaces.
Not to mention the long lines,
where people aren't even covering their faces?

Brad has become creative,
as he's building cool things in his room;

I'm not allowed in so I can't say what,
but it kinda smells like doom.
COMIX
STUFF
PLANS
THE CALCULUS 1
PAINT

I'm a little confused by this flu,
what is a patient zero?
My awesome mom has struggled a bit,
while Dad has become my new hero.

Dad understands me best,
he says this is like a temporary game.
"This too shall pass," he says,
we wish Mom felt the same.

I guess everyone's handling this,
the absolute best they can;
Getting through each day is tough,
Mr. Virus, I'm not a fan!

Each day is a brand-new adventure,
my emotions are everywhere;
Some days good, other days not,
it all seems a bit unfair.

I

Today it's been six weeks,
and the truth is finally settling in;
Sometimes Mom enjoys some wine,
while Dad his tonic & gin.

Don't we live in the 21st century,
why is everyone freaking out;
For toilet paper and paper towels,
what's that all about?

Mom feels better knowing
this flu has affected everyone;
She hopes it brings the world closer,
while I hope we can have more fun.

My Dad just smiles sweetly,
and tells me day after day;
Aside from the world-wide experience,
we each have our own way.

I

Dad says, "Unique as you and me,
to each exists our own path."
Emotions and feelings running amuck,
it will be one interesting aftermath.

My brother says, "Emotions are like farts,
they come and go as they please."
We all laugh and it breaks the tension.
Again, I wish time could freeze.

I

Life is pretty good for me
except I keep feeling that weight;
My Dad says it's natural, but what do I know?
I'm really only eight.

My family is home all the time,
my brother less annoying than normal;
I can't get used to scrubbing my hands this much,
it's so very abnormal.

The law says we must wear masks,
I pretend I'm a bunch of superheros;
My brother hates them with a passion,
when it fogs up his new subzeros.

Nothing will ever be the same again,
nothing the way it used to be;
We're all just doing our best to survive,
wouldn't you agree?

Today it's been about eight weeks,
welcome to our new 'normal'.
This is really really uncomfortable,
like a dance or semi-formal.

We have our family routines
with movie or game night Wednesdays;
The whole family laughs together
eating delicious pizza gourmets.

Twice last week I caught a glimpse
of a former Mom smile.
I kept to myself as not to disturb,
I'll admit, it's been a while.

Dad says to accept the world as-is,
and Mom will eventually come around;
Dad keeps preparing us daily saying,
"It's as uncertain as a battleground."

My family is a bit unique,
but I love them 'cause we have a ball;
Wacky and all, would be nowhere else,
for this once-in-a-lifetime close call.

Brad lets me into his room, sometimes,
I know it's not a done deal;
I'm working on him slowly,
perhaps one day, it'll be for real.

I'll tell you a secret of mine,
I've been feeling this very spring;
Sometimes I wanna run away,
sometimes I wouldn't change a thing.

I've learned a thing or two,
through this recent pandemic age;
We all have gifts and emotions to share
on our proud refrigerator stage.

In life, we all have bad days,
followed by good days indeed;
Lately, we struggle and fight,
instead of trying to succeed.

Today it's been two and a half months,
it's old and tired, I'll tell whoever;
Some days are longer than others,
but the City has changed forever.

I've never seen the flowers,
trees or bees quite so alive;
it appears without humans,
they can flourish and thrive.

Mom is super excited,
to return to work one day soon;
She finally apologized,
and appears to have changed her tune.

Mom feels she dealt with this poorly;
confused, angry, sad, amen;
But we'll all forgive her, surely,
Dad was right not to judge or condemn.

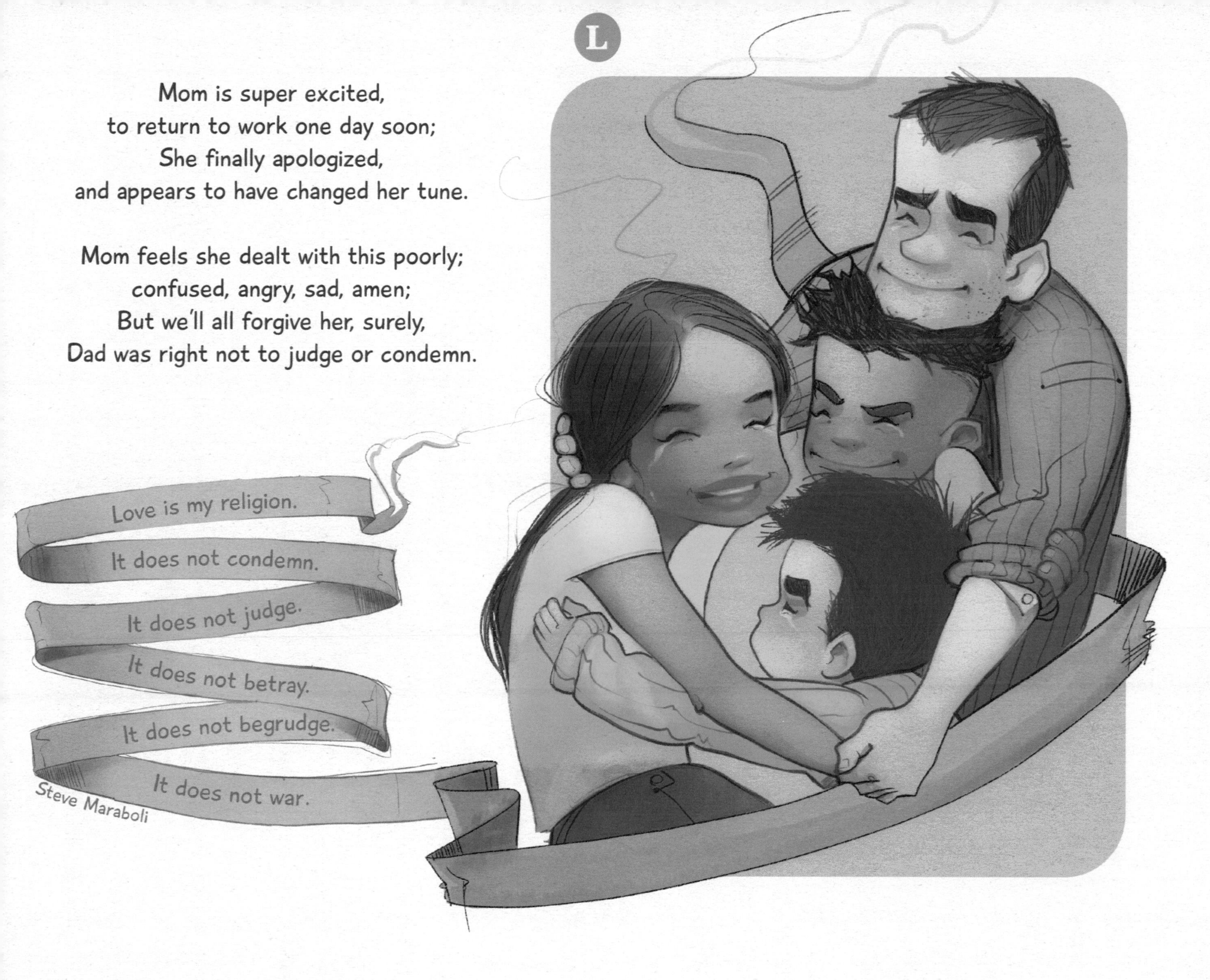

Dad says we can only do our best,
and that's all anyone can expect;
He says we'll have to wear masks,
for months and months, we checked!

Brad hates his mask with a passion,
he always goes a bit ballistic;
I love my mask, it's super cool
and a tad futuristic.

Today it's four months later...
and the new normal is our reality!
The City is slowly reopening
and the streets nearing vitality.

Wearing masks is the latest fashion trend,
toilet paper once again in stock;
Been grateful these last few weeks,
'cause we are now permitted to walk!

Mom is happier now, she says,
"Together, we can survive anything;
We will always help each other,
because family... means everything".

Dad says, "When we know better,
we do better and one day I'll see..."
No matter what happens, we'll get through it,
together, my family and me.

Dad talks about being grateful,
gracious, happy, and forgiving;
'cause it brought us even closer
and we are lucky to be living.

Life as we knew it changed,
life as we knew it ceased;
I'm grateful to be alive,
and the good in humanity released.

I've heard people talk of a cleanse,
nature's way of resetting the Earth;
Humanity is forever changed,
and I feel a universal re-birth.

E

Someday soon we'll play outside,
no masks, no worries, no fear.
Someday soon, we'll see my grandparents,
I'm hopeful within the year.

We live to enjoy another day,
we live to share our gifts.
The City's upside down now,
after protests and radical shifts.

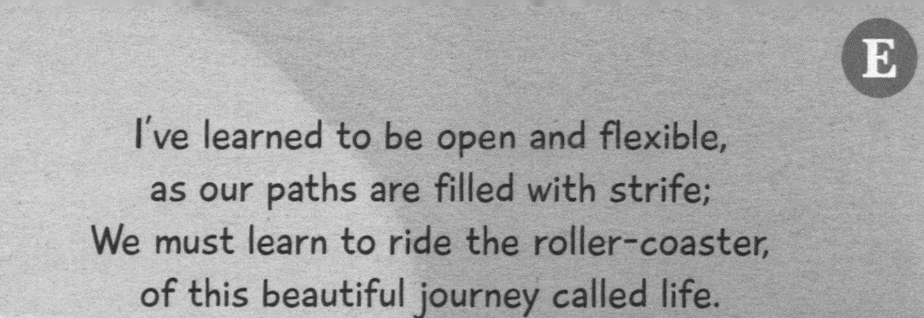

E

I've learned to be open and flexible,
as our paths are filled with strife;
We must learn to ride the roller-coaster,
of this beautiful journey called life.

And there you have it, my true story,
of this ferocious flu;
I hope you've learned through this crisis,
our loved ones are truly the glue.

Healing from loss is not easy,
there's surely no debate.
Get ready and buckle up,
'cause LIFE is what YOU create!

And I would say we lived happily ever after,
but we know this much is true...
In times of trauma and crises,
we uncover the real you!

S **SHOCK**
Nature's first line of defense to protect the mind, body, and soul.

M **MOCK-CCEPTANCE**
The pause between the painful event AND the pain itself.

I **IN-OVERWHELMDOM**
Frustration. Anger. Sadness. Tears of Laughter. Significant emotions.

L **LEARNING**
Now you know better, so you will do better!

E **EMBRACE**
Accept what has happened with grace, strength, and dignity.

The **S.M.I.L.E.** Adventure

Talk to your kids

S

SHOCK

a) Ask them how they are doing and what they are feeling
b) Share how YOU are doing and what YOU are experiencing

M

MOCK-CCEPTANCE

a) Talk about what's happening now
b) Tell them it's okay to feel anything and everything these days

I

IN-OVERWHELMDOM

a) Encourage them to share their thoughts and feelings with you
b) Share YOUR thoughts and feelings with them
c) Acknowledge & validate - be kind, patient, honest and transparent
d) Ask them to share when they're feeling icky and YOU do the same
e) Assure them we will all get through this

L

LEARNING

a) Ask them what they've learned thus far
b) Share what YOU've learned thus far
c) Discuss changes in families, school, present, and future
d) Assure them everything will be better in the new normal

EMBRACE

a) Inspire them to be creative and express themselves
b) Lead by example and be the vision you wish to create

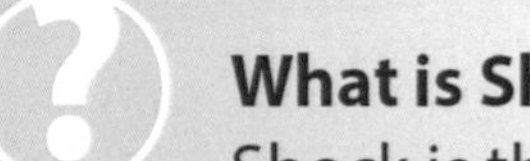

What is Shock?
Shock is the mind's way of saying, "Whoa! Wait a second! This cannot be real! Wow! Is this really happening?" Shock is a natural, physical and emotional reaction to surprising or upsetting news. Sometimes, it's just too much for us to handle.

Shock allows us the time and space we need to understand what's going on or what has just happened. How cool is it that? We, as humans, have a powerful automatic response system in order to protect and shield us from an emotional and physical overload (when it's just too much)?

Feelings you may experience!
You probably feel a bunch of different things right now. Everything might be going slow right now. You might be sad, angry, or upset. That's perfectly okay. You also might be feeling numb, nervous, scared, or afraid.

What you need to know?
This is completely natural. Keep breathing. You are not alone. Everyone goes through this. Take care of yourself and your family. Be gentle, patient and nice with yourself and everyone around you. Remember to give people around you a little space and time. Each person deals with the Shock differently. For some, it could last minutes. For others, it could probably last hours, days, or even weeks.

1. Sometimes you just need someone to sit with you in silence.
2. There is no wrong way to be or act right now!
3. There's nothing to worry about, until there's something to worry about.
4. It's okay not to feel anything! Or perhaps you are feeling everything!
5. Everything will be all right!

Use an example from the story or talk about how **THEY (or YOU)** have experienced **Shock** in the recent past?

Mock-cceptance

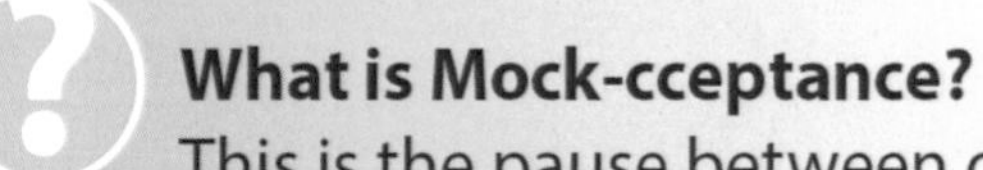

What is Mock-cceptance?

This is the pause between denial and acceptance. Sometimes when we have stress, hurt, or pain, it's too much for us and we are not quite ready to handle all the emotions and feelings.

Mock-cceptance is this safe place in the middle between the hurtful or painful event AND the feelings associated. Our body has created this incredible layer of protection which shields us from all the feelings.

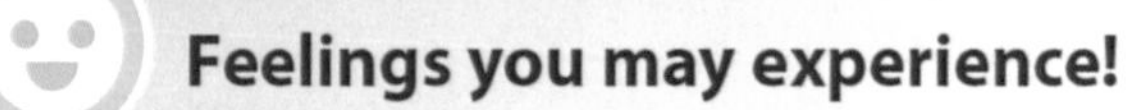

Feelings you may experience!

Sometimes you will feel good and sometimes you won't. Sometimes you might even feel numb. You will probably switch feelings from good to bad, and back and forth. It may even be sad at times. Any feelings you have are valid and okay.

What you need to know?

Feel your feelings whether that's numbness, anger, or sadness. It's important for you to do so. This is completely natural. Share with family members or people you trust. Keep breathing. Be gentle, patient and kind to yourself and everyone around you. Remember to give yourself and people around you a little space and time. Each person deals with Mock-cceptamce differently. This could last weeks or even months.

1. Ask for help from those around you.
2. It's okay to feel nothing or numb…even now.
3. Keep up your normal routine or create a new one!
4. No major decisions.
5. Be honest with your feelings.

Use an example from the story or talk about how **THEY (or YOU)** have experienced **Mock-cceptance** in the recent past?

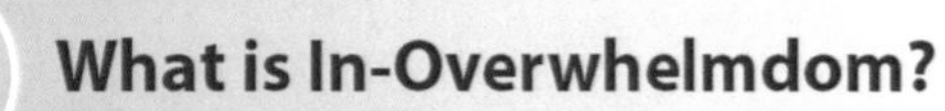

What is In-Overwhelmdom?

These are very intense and important emotions and feelings. Anger. Hurt. Sadness. Doubt. Worry. Numbness. The pain is here. The hurt is here. Your feelings will be anywhere and everywhere…and that's okay.

Feelings you may experience!

Here, you only have one choice. The only way through these feelings is to go through them. If you're angry, get angry. If you're sad, be sad and cry. Ride the wave of your feelings like a champ. You will cry. You will laugh. You will have very little patience. Here are other emotions you might feel: Anger, Hurt, Fear, Sadness, Happiness, Pain, Tired, Numbness, Discomfort, Afraid, Love, Hate, Nervous, Embarrassed, and tearful (for no reason and/or without warning).

What you need to know?

All of these emotions are valid, normal, and part of your adventure. Remember, emotions are like farts, they come and go. If you are sad, be sad. If you are angry, be angry. Then, do something else for a bit. Express your feelings to people you love and trust. It might be tough and it might get uncomfortable. These feelings will pass. There's no escape from this emotional roller-coaster, no matter how much it hurts to feel or how much it stinks. Everyone goes through this. It will get better and you will get over these feelings. Lastly, be gentle, patient and kind to yourself and others near you. Give yourself and those around you space and time to express, explore, and experience their feelings.

1. Trust the Process. Acknowledge the emotions and the crazinesses.
2. When one door closes, a new one opens.
3. Get comfortable. Relax. Watch something!
4. There are NO expectations, necessities, and/or requirements here. Everything is okay and will be okay.
5. Feel your feelings.

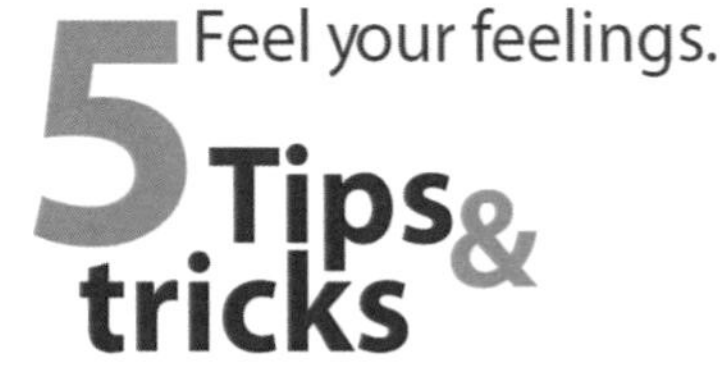

Use an example from the story or talk about how **THEY (or YOU)** have experienced **In-Overwhelmedom** in the recent past?

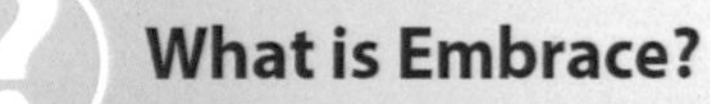

What is Embrace?

When you hug someone, you squeeze them close, tightly, to indicate you love them, forgive them, and are happy they are in your life, right? This is the same for everything we experience in life. When something painful or hurtful happens, you must eventually be forgiving, happy, enthusiastic and thankful for it. Otherwise, it could eat you up from the inside. Shock allows us the time and space we need to understand what's going on or what has just happened. How cool is it that? We, as humans, have a powerful automatic response system in order to protect and shield us from an emotional and physical overload (when it's just too much)?

Feelings you may experience!

Happiness. Smiles. Feeling alive. Feeling grateful. Enthusiasm. Passion. Energy. Peace. Love.

What you need to know?

It might not be as easy to Embrace the painful and hurtful events in your life as it is for the triumphs, the good stuff, and the fun and fantastic events. Either way, you appreciate that this event has happened and you understand that it's part of your history and is now a part of you. Therefore, to embrace is all about you. It's the celebration of you, who you are now, where you've come from, and where you are going. Moving toward a place of happiness or contentment, you must Embrace what has happened completel and move forward with your life. Finally, it has helped you to become the amazing and incredible person you are today. Be different. Be you! Shine!

5 Tips & tricks

1. Be grateful and forgive!
2. Remember the law Of attraction.
3. Celebrate who you are!
4. The best is yet to come!
5. Be unique. Be you at all times.

Use an example from the story or talk about how **THEY (or YOU)** have experienced **Embrace** in the recent past?

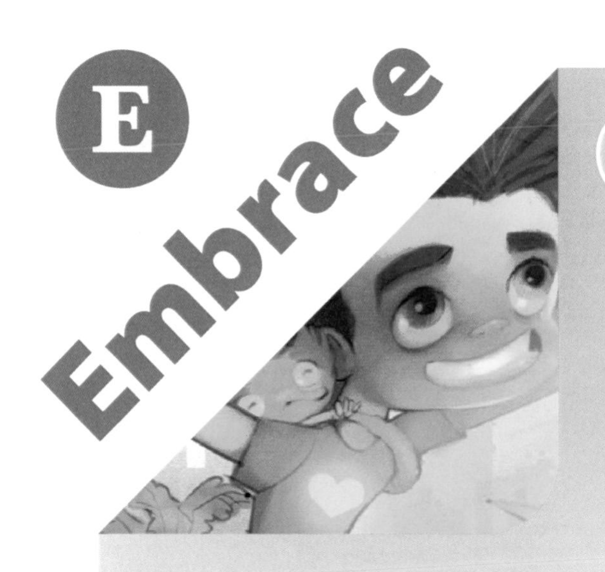

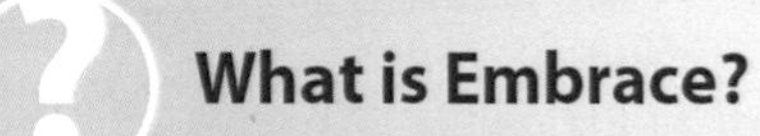

What is Embrace?

When you hug someone, you squeeze them close, tightly, to indicate you love them, forgive them, and are happy they are in your life, right? This is the same for everything we experience in life. When something painful or hurtful happens, you must eventually be forgiving, happy, enthusiastic and thankful for it. Otherwise, it could eat you up from the inside. Shock allows us the time and space we need to understand what's going on or what has just happened. How cool is it that? We, as humans, have a powerful automatic response system in order to protect and shield us from an emotional and physical overload (when it's just too much)?

Feelings you may experience!

Happiness. Smiles. Feeling alive. Feeling grateful. Enthusiasm. Passion. Energy. Peace. Love.

What you need to know?

It might not be as easy to Embrace the painful and hurtful events in your life as it is for the triumphs, the good stuff, and the fun and fantastic events. Either way, you appreciate that this event has happened and you understand that it's part of your history and is now a part of you. Therefore, to embrace is all about you. It's the celebration of you, who you are now, where you've come from, and where you are going. Moving toward a place of happiness or contentment, you must Embrace what has happened completely and move forward with your life. Finally, it has helped you to become the amazing and incredible person you are today. Be different. Be you! Shine!

1 Be grateful and forgive!

2 Remember the law Of attraction.

3 Celebrate who you are!

4 The best is yet to come!

5 Be unique. Be you at all times.

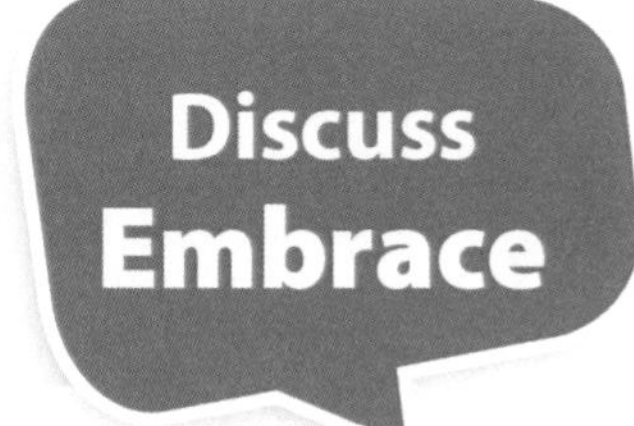

Use an example from the story or talk about how **THEY (or YOU)** have experienced **Embrace** in the recent past?

The end

CPSIA information can be obtained
at www.ICGtesting.com
Printed in the USA
LVRC081635100721
692202LV00018B/314

* 9 7 8 1 7 3 7 4 8 3 2 0 5 *